FLYING AND FLOATING

DAVID GLOVER

Kingfisher Books

NEW YORK

KINGFISHER BOOKS
Grisewood & Dempsey Inc.
95 Madison Avenue
New York, New York 10016

First American edition 1993
10 9 8 7 6 5 4 3 2 1 (lib. bdg.)
10 9 8 7 6 5 4 3 2 1 (pbk.)
© Grisewood & Dempsey Ltd. 1993

Library of Congress
Cataloging-in-Publication Data
Glover, David
    Flying and floating/David Glover —
1st American ed.
    p. cm. — (Young Discoverers)
Includes index.
    Summary: Investigates the properties of
air and water through experiments.
    1. Air — Juvenile literature. 2. Air —
Experiments — Juvenile literature.
3. Water — Juvenile literature. 4. Water
Experiments — Juvenile literature.
5. Evaporation — Juvenile literature.
6. Condensation — Juvenile literature.
[1. Air — Experiments. 2. Water —
Experiments. 3. Experiments .]
I. Title. II Series. QC161.2.G56 1993
533'.6'078—dc20 92-40212 CIP AC

ISBN 1-85697-843-5 (lib. bdg.)
ISBN 1-85697-937-7 (pbk.)

Series editor: Sue Nicholson
Designer: Ben White
Picture research: Elaine Willis
Cover design: Dave West
Cover illustration: Kuo Kang Chen
    Illustrators: Kuo Kang Chen pp.2, 7 (right), 9,
    11, 12-17, 18 (top), 19-26, 27 (bot.), 28 (left),
    29-31; Chris Forsey pp.4-5, 6, 13 (right); Kevin
    Maddison pp.7 (right), 8, 10, 18 (bot.), 27
    (top), 28 (top and right)
Photographs: Heather Angel p.25, Andrew Hill/
    Hutchison Library p.24; Christine Osbourne
    Pictures p.29; ZEFA p.19

Printed in Spain

About This Book

This book tells you about air and water – what they are and how we use their properties to fly planes or float boats. It also suggests lots of experiments and things to look out for.

You should be able to find nearly everything you need for the experiments around your home. You may need to buy some items, but they are all cheap and easy to find. Sometimes you will need to ask an adult to help you, such as when drilling holes.

Be a Smart Scientist

● Before you begin an experiment, read the instructions carefully and collect all the things you need.

● Put on some old clothes or wear a smock.

● When you have finished, clear everything away, especially sharp things like knives and scissors, and wash your hands.

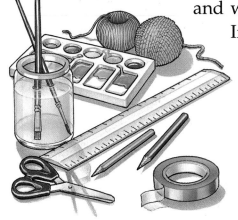

● Keep a record of what you do and what you find out. If your results aren't quite the same as those in this book don't worry. See if you can work out what has happened, and why.

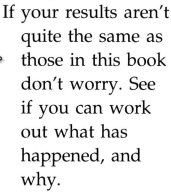

Contents

Air and Water

Air and water are essential for life on Earth. Without air to breathe and water to drink we could not survive. But what exactly are air and water?

Air and water have different properties — they look, feel, and behave in different ways. Air is a gas. You can't see it or taste it but you can feel it when it blows over your face and hands. Water is usually a liquid, but it can also be a solid or a gas (see page 18).

Without water and air, all animals and plants would die. Animals use the oxygen gas from the air when they breathe in. Much of the oxygen in the air is replaced by green plants, which produce oxygen in their leaves when they make food.

The Atmosphere

The atmosphere is a layer of air around the Earth. It is often divided into four bands — the troposphere (nearest Earth), stratosphere, ionosphere, and exosphere.

Air is mostly made up of the two gases nitrogen (78%) and oxygen (21%). It also contains carbon dioxide gas (0.03%), water in the form of a gas (water vapor), and tiny bits of salt, dust, and dirt.

After about 300 miles (500 km), the atmosphere starts to fade into space.

exosphere

ionosphere

stratosphere

troposphere

Do it yourself

Try these four simple experiments to compare the properties of air and water.

Blow up one balloon and fill another with water, keeping it over the sink so you don't make a mess. To fill the water balloon, stretch the balloon's neck over the end of the faucet, then slowly turn on the water. Or you could use a funnel. Knot the end of each balloon.

1. First, hold one balloon in each hand to see which is the heavier. You'll find that the balloon filled with water is much heavier than the one filled with air.

2. Now squeeze each balloon. The air balloon will be quite firm and taut and will feel a little springy. The one filled with water will change shape as the water sloshes around inside.

3. Next, try sinking the balloons in water. You will be able to push the water-filled balloon under the water more easily than the air-filled balloon.

Our Blue Planet

Over two-thirds of the Earth's surface is covered with water. Pictures of the whole Earth usually show the side with land, but from the other side our planet looks blue.

Pacific Ocean

4. Finally, stick a pin in each balloon. The air balloon will pop as the air rushes out, but the water balloon will simply sag as the water trickles away.

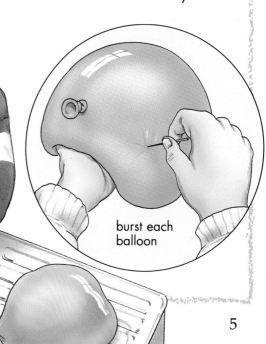

burst each balloon

The Weight of Air

Air has weight, even though it seems thin and light. For example, the air inside a bus weighs as much as one of the passengers! Air is one thousand times lighter than water. This means that a bathful of air weighs about the same as a cupful of water. Air also takes up space — see how air bubbles fill the bottle in the experiment below.

Blow It Up

When you pump up the tires on a bike, you're increasing the air pressure inside so that it can support both you and the bike (see next page).

Do it yourself

Here's a fun way to measure how much air your lungs hold (and find out more about the properties of air). You'll need a bowl, a large plastic bottle, and a long tube.

1. Pour water into the bowl so it is three-fourths full. Then completely fill the bottle with water. Put your fingers over the top of the bottle so the water doesn't spill, turn it upside down and lower it into the bowl.

2. Thread the tube into the neck of the bottle. Then take a deep breath and blow down the tube.

How It Works

Air is lighter than water, and it takes up space. As you blow, air bubbles rise up the tube and push the water out of the neck of the bottle into the bowl.

Can you fill the bottle with air?

The weight of all the air in the Earth's atmosphere pushes on everything with a force called air pressure. It may be hard to believe, but the air pressing on every square inch of your body weighs nearly 15 pounds! The reason you're not crushed by all that weight is because the air and liquids inside you push back with the same force.

Eye-Spy

Have you ever sucked air from a plastic bottle or a thin metal can when you're drinking from it? Air pressure flattens the sides of the bottle or can slightly. The sides pop out again when you let air back in.

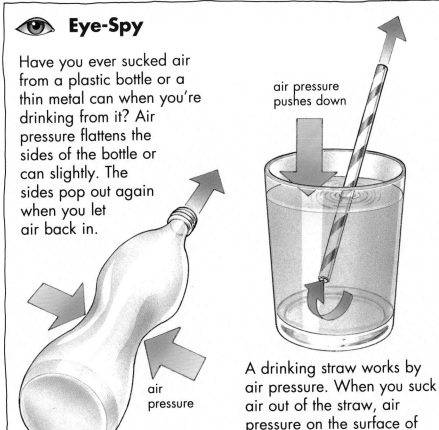

air pressure

air pressure pushes down

A drinking straw works by air pressure. When you suck air out of the straw, air pressure on the surface of the drink pushes the liquid up into your mouth.

Above: The amount of air in the atmosphere varies — the higher you go, the less there is. We describe this as the air getting thinner. Climbers of very high mountains, such as Mount Everest, must carry tanks of oxygen gas with them so they can still breathe.

Hot Air

Have you ever wondered why smoke from a fire goes up into the air rather than spreading out? The reason is that hot air is lighter than cold air, so hot air rises.

Rising hot air is used to lift huge air balloons. A hot-air balloon is simply a large bag made out of a light material. The air inside a hot-air balloon is heated by a gas burner. This makes the balloon float up into the cooler atmosphere.

Lighter Than Air

Some gases, such as helium, are lighter than air. This means that they don't need to be heated to help things fly. Helium-filled airships are used to film sports events such as the Olympic Games.

gas burner

Do it yourself

Make a hot-air pinwheel and balloon to see how hot air rises.

Hot-Air Pinwheel

1. Cut out a circle of cardboard and make eight cuts, as shown. Twist up the edges slightly (so that air can rise through the gaps).

2. Push a pin through the middle of the cardboard to make a hole. Thread string through the hole and knot the end. (Make a large knot so the string doesn't slip through.)

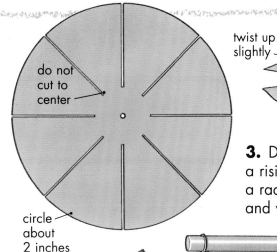

do not cut to center

circle about 2 inches across

twist up slightly

3. Dangle the pinwheel over a rising current of hot air (from a radiator, for example) — and watch it spin.

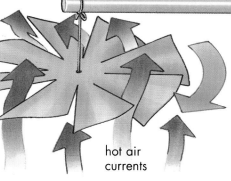

tie to a wooden stick or drinking straw

hot air currents

Hot-Air Balloon

1. Cut out eight 2-foot-long pieces of tissue paper to the shape shown.

2. Glue the edges together to make a balloon shape. Don't use too much glue or your balloon will be too heavy.

3. Fill the balloon with hot air from a hair dryer. Does it fly? You could make a small tissue paper box and tie it to the bottom of the balloon to see whether the balloon can carry a load.

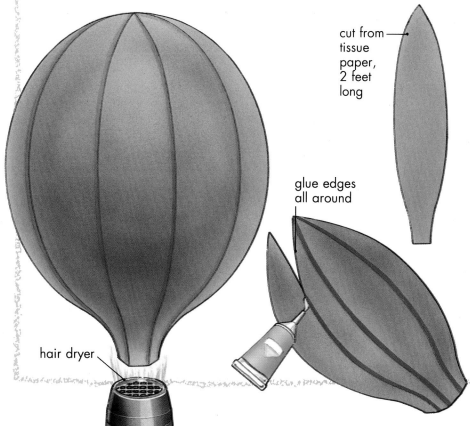

cut from tissue paper, 2 feet long

glue edges all around

hair dryer

The Force of the Wind

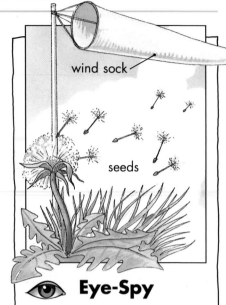

wind sock

seeds

When hot air rises, cool air usually flows in to take its place. This is what makes the wind blow. During the day, the Sun heats up parts of the land and the sea, which then warm the air above them. This warm air rises into the atmosphere and a stream of cooler air — the wind — blows in.

Winds carry seeds, make flags flap, and fill sails on boats. Strong winds can knock down trees and buildings.

👁 Eye-Spy

Look out for things that are moved by the push of the wind. Wind socks show which way the wind is blowing. Seeds are spread by the wind.

Beaufort Scale

In 1805, a British admiral called Sir Francis Beaufort introduced a scale to show the effect of different wind strengths on sailing ships at sea. Today, the Beaufort Scale describes the strength of the wind on land as well as at sea.

The scale has 12 numbers, ranging from calm to a violent hurricane.

Force 0

Calm

smoke goes straight up

Force 1-3

Light breeze

leaves rustle; small branches move

Force 4-5

Moderate wind

small trees sway

Force 6-7

Strong wind

large trees sway

Force 8-9

Gale

shingles fall off roofs

Force 10-11

Storm

widespread damage

Force 12

Hurricane

disaster

Most things are heavier than air, so they do not float upward on a still day. But on a windy day, the wind can carry some lightweight things into the sky. This is how a kite works — it's lifted by the force of the wind.

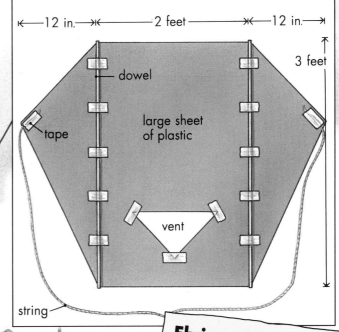

tie loop of string to main ball of string

Do it yourself

Make a kite! You'll need a large piece of strong plastic (cut from a bag), two light sticks or lengths of dowel, tape, and strong string.

1. Open out the plastic and cut it to the shape and size shown. Cut out a triangular hole (a vent) and strengthen the corners with tape.

2. Tape the sticks to the plastic and tie on the string.

|←—12 in.—→|←———2 feet———→|←—12 in.—→|

dowel

3 feet

tape

large sheet of plastic

vent

string

Flying Hints

First choose a windy day! Unwind your string about a yard. You may need to run into the breeze to start with. Let the string out as the kite starts to lift.

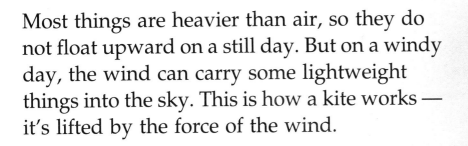

WARNING: Never fly your kite near overhead power lines.

Gliding Through the Air

Hot-air balloons and kites fly because they are lighter than air. But most aircraft are heavier than air. They need wings to help them to fly. Wings are a special shape — they're curved more above than below. They work when they move through the air at just the right angle. The flow of air above and below the wing makes a force called lift which keeps the aircraft in the air.

1. Many birds push themselves into the air by flapping their broad, flat wings. Some birds, such as gulls, soar and glide on moving air currents.

Using Thermals

Thermals are created when dark surfaces, such as roads and fields, soak up heat from the Sun then pass on the heat to the air above. Gliders and soaring birds use these rising warm air currents to fly higher and stay airborne longer.

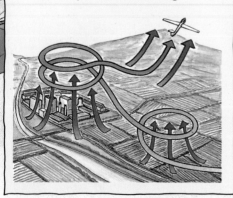

lightweight metal frame

pilot moves body to steer into a breeze

2. Hang gliders are usually launched by running down slopes or jumping off cliffs.

3. Gliders drift slowly downward at a gentle angle unless the pilot can find a rising air current to carry the lightweight glider up again.

When you cycle fast you can feel the air rushing past your body. But the faster you go the more the air drags you back. Modern cars and aircraft are smooth and rounded so that they pass through the air with as little drag, or air resistance, as possible.

Do it yourself

Check how the flow of air acts on a wing, making lift.

4. Parachutes fall rather than fly. The open parachute fills with air and works like a brake, using the air's drag to slow itself down.

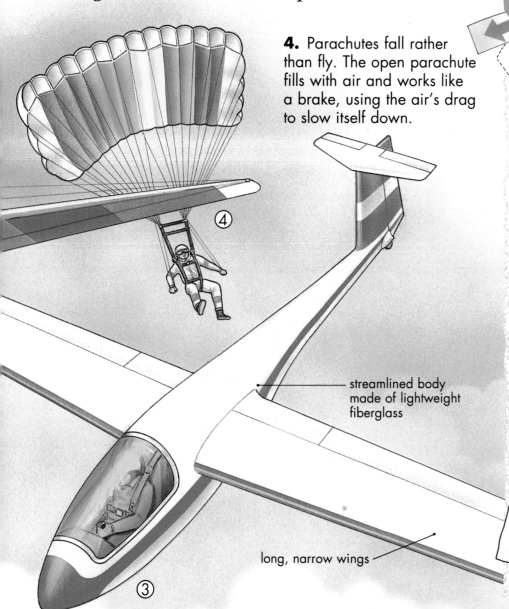

④

streamlined body made of lightweight fiberglass

long, narrow wings

③

Find a thin sheet of paper and hold it just under your lips. Blow hard over the paper and watch what happens.

You'll find that the paper will lift. The air you have blown over the paper moved faster and had less pressure than the air under the paper. This difference in air pressure made the paper lift.

Staying Up

All aircraft with wings will only fly if they can move fast enough to keep air flowing around their wings (making lift). They also need enough lift to overcome their weight.

Do it yourself

Make a parachute!

1. Cut a square of plastic from a plastic bag or some fine, lightweight material. Make a small hole in the center. (This makes the parachute more stable.)

2. Tape a thread to each corner of the chute.

3. Tie the free ends of the threads together and tape the knot to a toy figure. Roll the chute into a ball and toss it into the air.

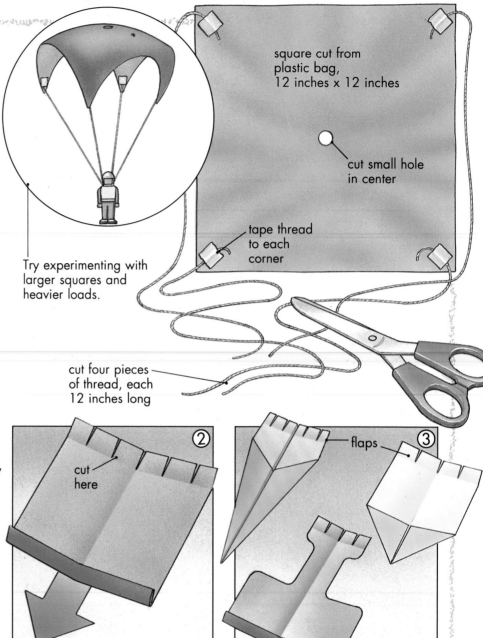

Try experimenting with larger squares and heavier loads.

square cut from plastic bag, 12 inches x 12 inches

cut small hole in center

tape thread to each corner

cut four pieces of thread, each 12 inches long

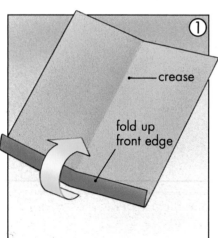

crease

fold up front edge

①

cut here

②

flaps

③

Fly a superglider!

1. Fold a sheet of rectangular paper in half, making a neat, stiff crease down the middle. Fold up the front edge two or three times.

2. Fold the back edge once and make a series of cuts (for (flaps), as shown. Launch the glider into the air. If it falls too fast, unfold some of the paper from the front. If it stalls (tips up at the front

then dips down), add more folds at the front. Bend the flaps to see how they change the glider's flight.

3. Try some of these glider shapes or design your own.

Planes and Helicopters

Gliders and hang gliders must go where air currents take them, but aircraft with engines can travel where the pilot wishes, and they can keep going until they run out of fuel. Engines keep an airplane pushing through the air so that its wings are working, creating lift.

Helicopters are different because they can fly backward, forward, and up and down. They create lift by pushing down on the air. Tilting the blades moves the helicopter in different directions.

Because they can hover, helicopters are useful for watching traffic, lifting loads, or rescuing people from the sea or from mountainsides.

Do it yourself

Make a helicopter rotor.

1. Carefully cut a rectangular strip from a plastic bottle.

2. Cut two slits, one on each side of the strip, as shown.

3. Make a small hole in the middle of the strip and push a pencil through so that it's a tight fit.

4. Hold the pencil between the palms of your hands, spin it quickly and let go.

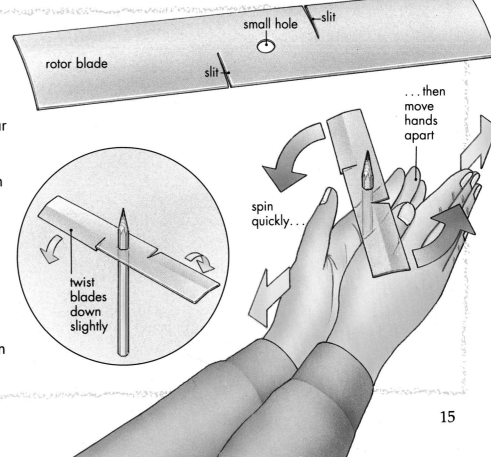

rotor blade

small hole

slit

slit

twist blades down slightly

spin quickly. . .

. . . then move hands apart

15

Wind Power

Did you know that there is more energy in a gale than a city uses in a year? For hundreds of years, people have used the power of the wind to sail ships or to turn windmill sails to grind corn. Wind power can be used to make electricity, too. Unlike coal and oil, wind power will never run out, so using it will help save these fuels.

Sailing Ships

The huge square sails on this old sailing ship were designed to catch the wind coming from behind. This is called sailing before the wind.

Do it yourself

Experiment with sails.

Make three or four model boats from Styrofoam to the design shown below.

Cut out different sail shapes and attach them to the masts. Float the boats on water and blow them along with a straw to see which one has the best sail.

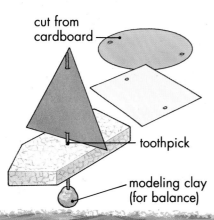

cut from cardboard

toothpick

modeling clay (for balance)

① ②

Do it yourself

Make a pinwheel windmill with a square of thin cardboard, a straw, a pin, and glue.

1. Make four cuts in the cardboard, as shown.

2. Fold over the edges and glue them in place.

3. Cut a short piece off the end of the straw. Push a pin through the center of the cardboard, short straw, and the full-length straw. Ask an adult to bend the back of the pin over.

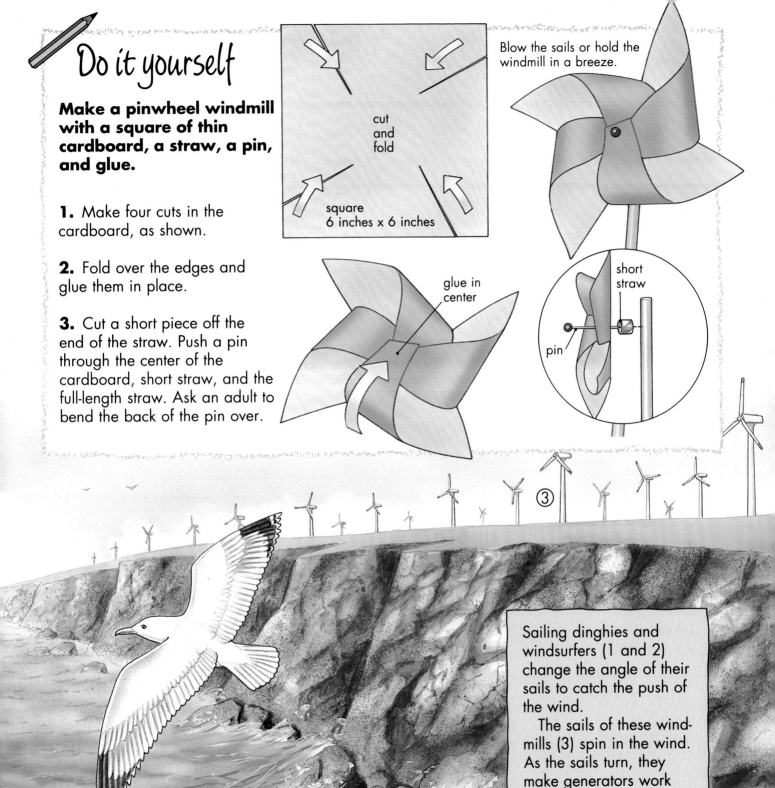

cut and fold

square
6 inches x 6 inches

glue in center

Blow the sails or hold the windmill in a breeze.

short straw

pin

③

Sailing dinghies and windsurfers (1 and 2) change the angle of their sails to catch the push of the wind.

The sails of these windmills (3) spin in the wind. As the sails turn, they make generators work to produce electricity.

A Watery World

We drink it, we wash with it, we swim in it — water is very special. Without water, our Earth would be a dry, lifeless desert. Water can exist in three different forms — as a liquid, as a gas (called water vapor), and as a solid (ice), and we use all three forms of water in our everyday lives. For example, steam irons use the gas from boiling water to press clothes, and we have iced drinks to keep cool.

gas

liquid

solid

Water freezes at 32°F (0°C) and boils at 212°F (100°C). Steam is a cloud of water droplets. Steam spurts from a boiling kettle because it takes up more space than the liquid.

Frozen Solid

Solid water forms in two different ways. Ice is made when liquid water cools to 32°F (0°C). This temperature is called water's "freezing point."

The second way is when water vapor freezes to make frost. For example, dew may freeze into frost on cold mornings.

clouds are water vapor

snow is frozen water

water vapor turns into steam in cold air

ice is frozen water

18

Gravity, the invisible force that pulls everything toward the center of the Earth, pulls on water too. In places where it cannot flow through surface rocks, water forms rivers, lakes, and oceans. In other places, different kinds of rock let the water seep through. All water on Earth eventually settles at the lowest level that it can possibly reach.

This photograph shows a waterfall in Ethiopia, Africa.

Do it yourself

See how the surface of water in any container is always level, no matter which way you tip it up. You'll need two clear plastic bottles with tops and a plastic tube.

1. Fill one of the bottles with water and screw on the top.

2. Connect the bottles with the plastic tube (like the kind used in wine-making). Use shampoo or ketchup bottles with nozzles, over which you can push the tube, or ask an adult to make holes through the bottle tops for you.

3. Turn the bottles upside down. Make a small hole in the bottom of each bottle to let the air in and out. Now change the height of the bottles and compare the water levels.

How It Works

You'll find that no matter where you hold the bottles, the water will flow from one to the other until the level of water is the same in them both.

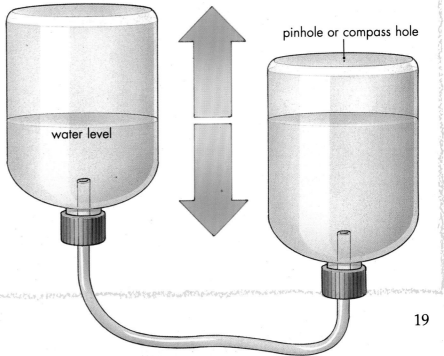

water level

pinhole or compass hole

19

Water Pressure

Can you swim under water? If so, perhaps you have felt pressure in your ears as you swim near the bottom of the pool. This pressure is caused by the weight of the water above you, pushing against you and into your ears. At the bottom of deep oceans, the water pressure is so enormous it would crush your body. Yet there are some strange creatures that can survive there.

Scuba divers can dive safely to about 300 feet (100 m).

Life in the Deep

The Sun's light and heat cannot reach deep down in the oceans, so it's dark and cold there. The deepsea angler fish carries its own small light in front of its mouth to attract its prey.

Some deepsea fish are actually held together by water pressure and most die when they are brought up from the deep.

Some strong steel submersibles, like Alvin, are specially designed so that they can dive to depths of 13,000 feet (4,000 m) and explore the seabed.

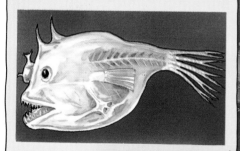

ALVIN

Do it yourself

If your "diver" has a hole in the top, fill it with clay to make it watertight.

Put a strip of tape over the holes, then peel it off when the bottle is full.

Make a bubble diver. You'll need a clear plastic bottle with a screw top, a plastic pen top, and some modeling clay.

1. Stick a small piece of modeling clay to the long stick part of the pen top. This is your diver.

   Test the diver in a glass of water to make sure that it floats properly. The diver should just float at the water's surface, as shown on the right. If it floats too high, add some more clay. If it sinks, remove some clay.

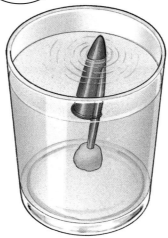

2. Fill the bottle with water right up to the brim. Drop the diver in the bottle and screw on the top.

👁 **Eye-Spy**

Cut about five holes in the side of a plastic bottle then fill the bottle with water. See how water pressure makes the water from the bottom hole squirt out the farthest.

3. To make it dive, squeeze the bottle. To make it surface, release the pressure.

How It Works

When you lower the diver into the water a bubble of air gets trapped inside the pen top. Squeezing the bottle increases the water pressure, making the air bubble smaller, and the diver sinks. When you let go, the bubble gets bigger again and the diver rises.

Floating and Sinking

Have you ever wondered why some things float on water while other things sink? A plastic ball will float, for example, but a stone will sink at once.

Whether something floats or sinks depends on its density (how heavy it is for its size), and its shape. The shape of an object controls how much water it pushes out of the way, or displaces. A plastic ball floats because it is lighter than water. It cannot displace much water with its weight. A stone is heavier than water so its weight pulls it under.

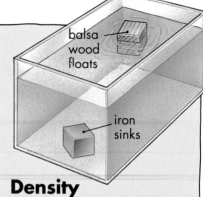

balsa wood floats

iron sinks

Density

A cube of iron is 7 times heavier than a cube of water the same size. But a cube of balsa wood is 5 times lighter than the water. If something is heavy for its size we say it has a high density. If it is light for its size, it has a low density.

👁 Eye-Spy

Next time you have a bath, see how far the water level rises as you lower your body into the water. This will show you how much water you displace!

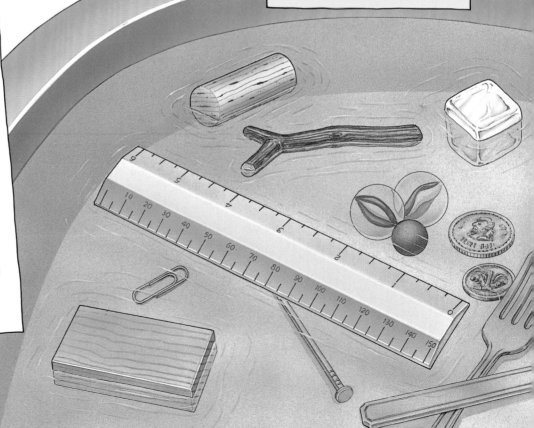

Do it yourself

Experiment with different materials to see whether they float or sink. Then try changing sinkers to floaters.

First test a collection of solid objects, such as things made of metal, wood, glass and plastic.

You'll find that small, heavy objects, such as coins and marbles, will sink. Larger, light things, such as a stick or a cork, will float.

Next, scrunch up a sheet of aluminum foil into a tight ball and drop it into the water. The ball will sink because it's heavier than the water it displaces.

Now fold another sheet of foil into a boat shape. Does it float or sink now?

Try the same experiment with a ball of modeling clay.

How It Works

The wide, flat boat shape displaces more water than the solid ball made from the same material. Most of the boat is actually filled with air. Together, the boat and the air inside it are lighter than the water they displace and so the boat floats.

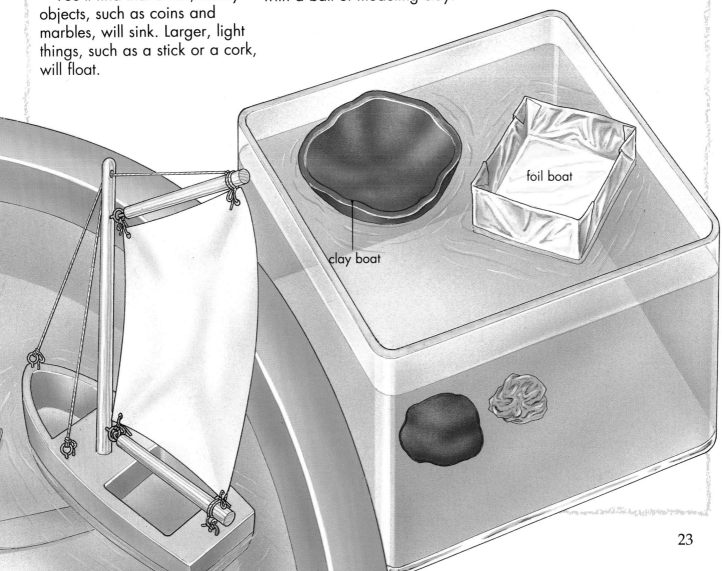

clay boat

foil boat

23

Sometimes boats capsize — they turn on their sides and fill with water. To reduce this risk, a boat must be made to the right shape and its cargo must be loaded correctly so that the boat is balanced.

Container ships like the one on the right are designed to carry heavy loads.

Do it yourself

See which boat designs are best at carrying cargo safely.

foil tray

marbles

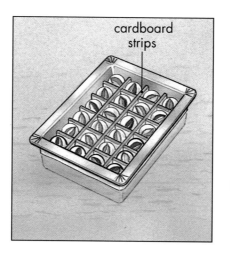

cardboard strips

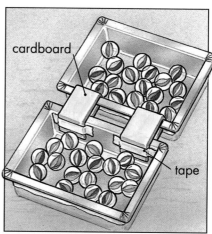

cardboard

tape

1. Float a foil dish on some water. Add some marbles one at a time for the cargo. You'll find that the marbles roll around in the boat until they tip it over altogether.

2. Divide up the inside with strips of cardboard to make the cargo more stable. Large container ships, like the one in the photograph above, are divided in a similar way.

3. Tape two foil trays side by side and balance the cargo again. This boat design is called a catamaran. It is a very stable design and will not capsize easily.

A Stretchy Skin

Have you ever noticed how water collects into tiny drops on a polished car or on shiny shoes? The weight of the water tries to make it flow and spread, but water has a kind of stretchy elastic skin that holds it together.

Fill a glass right up to the top with water. Can you see how the water's surface bulges slightly over the brim of the glass? The force holding the water together is surface tension.

Some lightweight animals, such as this spider, use surface tension to walk on water.

Do it yourself

A metal needle has a higher density than water and should sink, but you can make it float on water's stretchy skin.

Slowly slide the needle across the rim of a filled glass, onto the water's surface. Be careful not to break the surface tension. If you find this hard, try floating the needle on a raft of tissue paper. The paper will become waterlogged and sink, leaving the needle floating on the water.

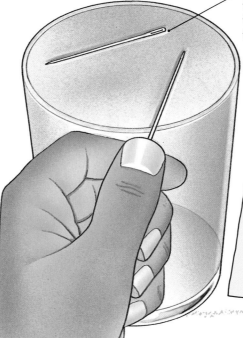

Look closely — you'll be able to see how the water's skin flexes under the weight of the needle.

More Things to Try

Add one or two drops of liquid soap to the water. You'll find that the needle will soon drop to the bottom of the glass as the soap weakens the water's surface tension. This property of soap is used in the experiment on the next page.

25

Do it yourself

Make a soap-powered boat. You will need some balsa wood and a tiny piece of hard soap.

1. Cut out the balsa wood to the shape shown in the circled drawing.

2. Fill a bowl with water, making sure that the water and the bowl are clean.

3. Push a small piece of soap into the triangular notch at the back of the boat and see how the boat is pulled forward.

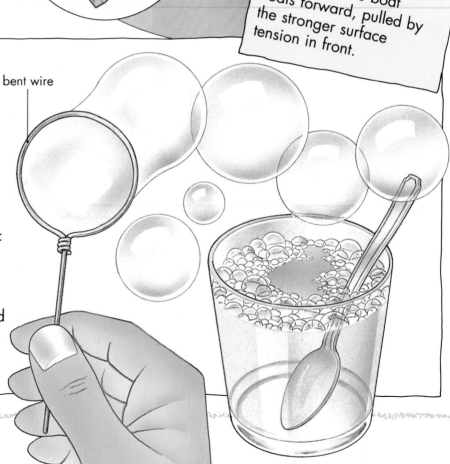

Change the water if you want to do this experiment several times.

soap

boat shape

How It Works

As the soap weakens the surface tension at the back of the boat, the boat floats forward, pulled by the stronger surface tension in front.

Blowing Bubbles

By lowering water's surface tension, soap can make it even more stretchy. That's why you can blow bubbles with soapy water. Here's a good recipe for soap bubbles.

bent wire

- Stir four tablespoonfuls of soap flakes into four cupfuls of warm water, making sure that the flakes dissolve.
- Leave the mixture to stand until cool.
- Stir in one teaspoonful of white sugar.

Puddles and Clouds

Next time it rains, wait for the rain to stop, then go out and see how long it takes for the rain puddles to disappear. Some of the water in puddles soaks down into the ground, but the rest dries up into the air. The Sun's heat makes the water turn into the gas we call water vapor. This process is called evaporation.

Water evaporates more quickly when there is a breeze blowing than when the air is still. That's why laundry hung outside dries faster on a windy day than a still one.

You can sometimes see water vapor rising from a damp dog as it lies steaming in the sunshine on a hot day.

Do it yourself

Investigate how quickly water evaporates.

Find three cloths of equal size. Soak them in water, then leave them in different places to dry:

1. Hang the first on a line.

2. Leave the second screwed up in a bowl.

3. Lay the third flat on the ground.

① Which one dries the fastest?

②

③

How It Works

The towel on the line will dry the fastest because the water vapor can escape from it more easily. If there's a breeze to carry the vapor away, the towel will dry even more quickly.

27

Condensation

Water vapor in the air can turn back into liquid water, too. This happens when warm air cools down. As cold air cannot hold as much water vapor as warm air, some of the water vapor forms tiny drops of liquid. This process of gas turning back into liquid is called condensation.

There is water vapor in the air you breathe out. If you breathe onto a cold glass mirror, some of the vapor in your breath condenses into a mist of tiny water droplets.

Eye-Spy

Watch out for water vapor condensing. You can often see it on the cool windows of a warm room, on the mirror of a steamy bathroom, or on the inside of a greenhouse. Plants give off water vapor at night (with carbon dioxide gas), which condenses on the greenhouse's cool glass.

Do it yourself

Leave some glasses in a freezer for about two minutes. When you bring them out, see how water vapor from the air condenses on their sides.

Below: Clouds are made from water vapor. When warm air rises, it eventually cools and the water vapor condenses into a mist of droplets. If the droplets get large enough, they fall from the sky and it rains.

Water Power

We can use the power of moving water just as we can use wind power. Turning water wheels were once used to grind corn or to drive tools such as drills. Today, water power is used mostly to make electricity. Water is stored in huge reservoirs behind dams. When it is released, the force of the falling water drives turbines which produce electrical power.

Water Wheels

Water wheels like this one were first built over two hundred years ago. This photograph shows a working wheel in Syria.

Do it yourself

Build a water wheel.

1. Ask an adult to help you make a hole through the middle of a cork and cut eight slots in its sides.

2. Glue a plastic spoon in each slot. When the glue is dry, slide a knitting needle through the hole in the cork.

3. To work the wheel, hold it under a stream of gently running water.

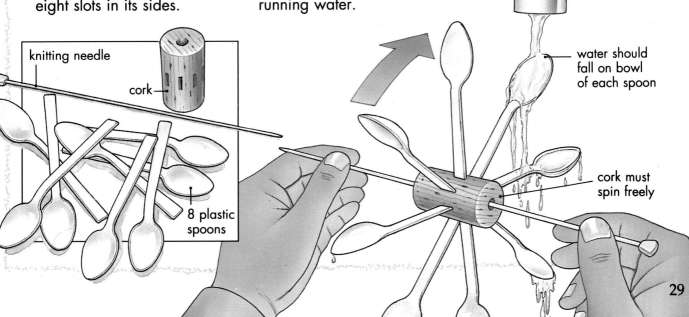

knitting needle

cork

8 plastic spoons

water should fall on bowl of each spoon

cork must spin freely

29

Clean Air, Clean Water

We must have clean air to breathe and clean water to drink to stay healthy. Pollution is the dirtying of our land, air, and water. Air pollution comes from dangerous gases given out by factory chimneys and by car and aircraft exhausts. These gases mix with the water vapor in the air to make acid rain. Acid rain is harming our land and water, and poisoning plants and animals.

carbon dioxide

Greenhouse Effect

This is a warming of the Earth's climate by some gases that act like the glass windows of a greenhouse, trapping heat inside the atmosphere. Carbon dioxide is a greenhouse gas.

Do it yourself

Test for air pollution around your home.

1. Find three shiny metal tin lids or small mirrors. These will be your dirt traps. Polish them with a clean dry cloth, then smear a little petroleum jelly in the center of each.

2. Leave each dirt trap in a different place. For example, leave one inside, in a living room or a bedroom; one in a windowbox or a yard; and one near a busy road.

3. After a few days, collect your dirt traps and compare the amount of dirt they have collected. Which is the most polluted place?

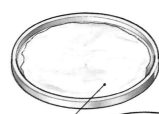

smear with petroleum jelly

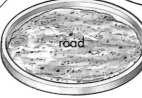

road

yard

room

What You Can Do

Here are some ways of making the air cleaner for yourself and for others:

- Never smoke.
- Don't burn garden trash — make it into compost.
- Buy products that are environment-friendly.
- Travel by bike for short journeys, rather than by bus or car.

The water we drink is filtered and cleaned at water treatment plants before we use it. But much of the Earth's water gets polluted by factory waste, by sewage, and by all the garbage that careless people dump in our oceans, rivers, and lakes.

Do it yourself

Make a water filter.

1. Soak soil and leaves in some water to make it muddy then pour it through a sieve.

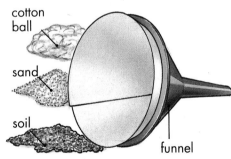

cotton ball

sand

soil

funnel

2. Put a filter paper in a funnel. Place a small cotton ball in the neck, then add some fine, clean sand.

dirty water

filter paper (or use thick tissue paper)

filtered water

3. Pour in some of your muddy water and see how the filter helps to trap some of the dirt.

👁 Eye-Spy

Take a look at any ponds and rivers near your home — which picture are they most like?

Help the environment by never throwing garbage into lakes and streams.

How It Works

The larger bits of dirt get trapped in the sand, cotton ball, and filter paper, letting cleaner water pass through. Don't drink it though — it's still very dirty!

Index

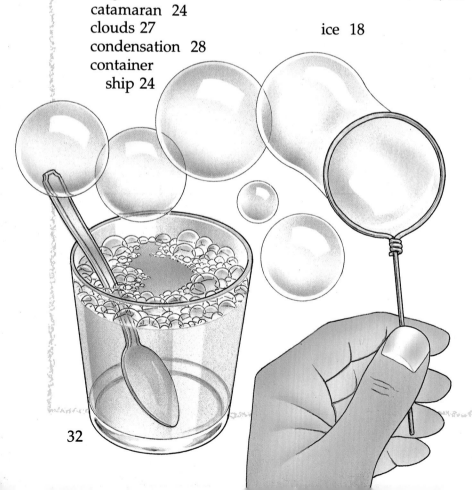